Spin-off Products and Services:

The Commercialization of Internally Supported Resources

by E. Patrick McGuire

A Research Report from The Conference Board's Division of Management Research Harold Stieglitz, Vice President

Contents

The Conference Board acknowledges with thanks the cooperation of those companies whose practices are described herein. Research for this study was carried out by E. Patrick McGuire, Project Director, under the direction of Earl L. Bailey, Director of Marketing Management Research.

Part I
Spin - off Ventures: Principles and Practices

Chapter 1
Introduction

To FUNCTION AT all, every company requires many different kinds of goods and services for its operations. Some of these are apt to be purchased from outside suppliers. But for a variety of reasons, a company may elect to be partly self-sustaining by producing internally certain of the items needed for its own use.

In some cases, such a company will later decide to offer an internally developed and supported product or service for routine commercial sale to other companies. The resulting commercial venture represents what may be called "spin-off" business for the company.[1]

Characteristics of Spin-off Units

An investigation of more than 152 spin-off situations, on which this report is based, found that the service or product involved had, in most instances, served a company's internal purposes for several years before being offered to others. A majority of the spin-off units studied had been established during the preceding decade.[2]

Furthermore, most of the operations studied remain relatively small. Few spin-offs are major contributors to the overall sales or profits of their sponsoring firms. While some are fully organized operating entities, many more rely heavily on the part-time contributions of personnel whose primary responsibilities lie elsewhere in their companies.

Spin-off Services

Services, rather than products, turn out to be the most common of all spin-off offerings. Some common kinds include:

- *Training and education services.* In several cases, services (and, sometimes, related products) developed originally to upgrade internal company skills – of groups ranging from top management to production mechanics to salesmen – are now being sold to outsiders. Included are in-house seminars, courses, programmed texts, audiovisual materials, etc.
- *Product or process design and engineering services.* Similarly, technical skills used in designing or making various kinds of equipment and other items for companies' own use are also being made available and sold to other companies.
- *Marketing research and analytical services.* After developing their own specialized techniques and services for "in-house" marketing research purposes – such as interviewing, panel testing, product use testing, methods for evaluating advertising or new products, etc. – some marketers now sell these to other firms as well.
- *Physical and chemical testing services.* Various testing services, once employed solely within their companies (often those relating particularly to pollution abatement and to occupational health and safety), are being offered by several spin-off groups.
- *Data processing services.* Some firms have found it worthwhile to establish service bureaus to extend their data processing services – facilities, staff, time-sharing programs – to others.

[1] "Spin-off" has other connotations, of course, in financial usage – as in describing the divestiture of the assets of a business. In the context of this report, however, the term "spin-off" refers solely to ventures in which products or services now sold partly to outsiders were originally developed merely for their own use by the sponsors.

[2] As reported later, not all spin-off ventures are set up as separate organizational entities. For convenience, however, such ventures are usually described as if they were.

Personnel services. Also, a number of companies surveyed are selling internally supported skills, techniques or materials which they originally used in recruiting, testing, motivating or evaluating their own employees.

Management consulting services. In still other cases, the services of captive management consultants now being made available to outsiders form the basis for spin-off units.

Spin-off Products

Where spin-offs offer products instead of services, these can range from raw materials to finished products. Many, however, are semi-finished components. Often, raw material surpluses are more likely to be marketed by full-fledged divisions or subsidiaries, rather than by spin-off groups. On the other hand, a company that manufactures certain of its own operating supplies, such as printing inks, cleaning materials, etc., may in time elect to sell them to others as well through a spin-off unit.

A few firms have employed a spin-off approach that makes use of specialized production machinery. For example, the Borden Company has a plant that produces can-closing machinery. This operation, originally set up to serve the company's own needs, now sells can-fabricating equipment and testing instruments to other food processors. Similarly, analytical instruments first designed for internal testing purposes of E.I. du Pont de Nemours & Company were later among the company's products sold by a spin-off unit.

Still other spin-off offerings combine both a product and a service. For example, an aircraft manufacturer developed specialized electroplating skills which it decided could be sold advantageously to other firms. Its spin-off group now provides not only electroplating services but also a variety of plating solutions. In another case, a manufacturer making its own printed circuits now supplies them also to nearby users in the electronics industry. The company's spin-off offerings include not only the time and skills of the group that prepares and processes its printed circuits, but also a special laminate used to manufacture circuit boards.

Chapter 2
Why Companies Set Up Spin-off Operations

IN EXPLAINING what led their firms to market spin-off products or services, managements cite various reasons. Generally, however, they have one or more of the following in mind:

(1) To make more efficient use of existing facilities or operations;
(2) To generate new revenues;
(3) To justify further having in-house capabilities for meeting the company's needs for specialized products or services;
(4) To provide opportunities for developing company personnel and giving them managerial experience.

Economies of Scale

The possibility for making better use of a company's production facilities or other resources is perhaps the single most common rationale for making internally developed items available to outsiders. Numerous firms report that by reducing unproductive "idle" or standby time – for production or service operations – they are often able to achieve significant economies of scale, thereby reducing their own costs of operation.

For example, a company running only a single shift may find it possible to make after-hours use of specialized production equipment. One such firm had purchased specialty spot-welding equipment for use in its own fabrication of sheet steel. Nearby smaller companies that lacked the funds to buy such equipment expressed interest in employing the welding facilities; and so a second shift of personnel was put on to handle custom-welding orders from these companies.

In another case, a chemical producer had installed a drum-washing line to recycle drums used to package the company's specialty chemicals. However, the washer was actually needed only one or two hours per day. Discussions with local drum reconditioners resulted in an arrangement whereby the drums from other companies in the area were first shipped to the chemicals manufacturer, processed through the washing line, and then sent to one of the reconditioners for final finishing. Fuller use of the washing line permitted the chemical manufacturer to cut the costs of washing its own drums by at least one-half.

Similarly, the costs of in-house services have also been reduced in many cases by selling such services to outside buyers. A "make or buy" decision regarding some internally required services, resolved in favor of the choice to "make," creates additional fixed costs. For some firms, these have become burdensome, especially during periods of economic stress, making spin-off prospects tempting.

An example is the data processing department of an appliance firm, which was in danger of being severely trimmed as part of a cost-cutting campaign. The department's manager, reluctant to lose key members of the unit's staff, devised a plan to sell computer services that made use of the firm's inventory control program to other companies. This "outside" work justified retaining the staff and prevented layoffs.

New Revenue Considerations

A related reason why many firms take their products or services "public" is the capacity of spin-off ventures to generate revenue – thus offsetting costs, or in some cases actually producing a profit. And while few of the companies studied have found their spin-off ventures to be financial bonanzas, many do report that their units are profitable.

Though fairly certain of bringing in sufficient revenues to offset at least part of the operational costs of their spin-off units, several managements were unsure at the outset of the exact sales and profit levels that could be attained from the ventures. Some still are not precisely sure how things have worked out. One metals company executive says, "It's pretty hard to know exactly where you stand as far as the 'profitability' of this operation. We do know that we're getting revenue that wouldn't otherwise be there. And that's enough reason for selling this service."

As reported by others, an executive in another firm states that the "kinds" of funds obtainable from the sale of a spin-off product were a consideration in the decision to proceed. He says that the income is regarded as "found" money – i.e., windfall revenue the company would otherwise not have.

"Making a profit" is seldom the primary mission, in any case. A number of firms report that they would have been satisfied simply to "break even" on their ventures. They note that other benefits derived from the formation and operation of spin-offs were even more important. "We're not looking to make a lot of money on this product," the spokesman for a furniture company explains. "We're happy that we can keep this operation going – serving our own needs – and also make some money besides."

Maintenance of Specialized Resources

A critical issue, in some cases, is that the revenues generated by the sale of spin-off products or services enables a company to maintain in-house production or service facilities that it had learned it could not otherwise afford. A few companies report that, without the prospects of outside sale, they would not have found it economically feasible to continue to draw upon their in-house operations.

For example, one reporting firm operates a sophisticated audiovisual arts department whose equipment was acquired in connection with a personnel education program developed some years ago. That program proved to be a one-time effort; once it was completed, the audiovisual equipment might have been sold. By renting the equipment to other firms, however, the company has been able to justify its retention.

The genesis of some spin-off projects goes back to past materials shortages, when companies have – in self-protection – elected to make products they would otherwise have purchased. Although the supply situation eased in time, these companies have been reluctant to lose their self-sufficiency. Again, spin-off sales offered an answer.

In still other cases, production economics does not permit companies to produce only the quantities of intermediate products needed for their finished goods. Instead of maintaining investment in inventories drawn from large-scale batches, these manufacturers have chosen to establish spin-off groups to supply these intermediates to other users.

Selling to outsiders, it is sometimes reported, affects not only the availability of an internally developed product or service, but its quality as well. Some firms say they have been forced to make quality improvements to meet the more demanding requirements of the marketplace. A chemical company, for example, began to market certain types of analytical instruments that it had designed and manufactured for its own use. It was discovered that, in order to satisfy commercial buyers, several design modifications were required to make the equipment easier to calibrate and use. Once these improvements were made, the firm's own laboratories became a beneficiary as well.

Management Development

A spin-off unit, according to a few companies, is sometimes a good place to test out management talent – to give promising candidates for higher level management posts a chance to show what they can do – or to experiment with new managerial techniques. While no firms report setting up spin-off units solely for these reasons, several do point out that such considerations contributed at least partially to their establishment of spin-off operations.

The spin-off unit is often small enough to provide an opportunity for checking out such approaches as new methods of motivating employees, or new-product promotion, without incurring major risks. And, if the commercial success of the unit hinges largely on the initiative and skills of its manager, senior management can observe how that manager performs.

A chemical company is one that finds spin-off units a useful "training ground" for its personnel. "Many of our middle managers are specialists," says a vice president of the company. "They are skilled in one or two areas of management – production, sales, etc. But they haven't had an opportunity to manage a 'complete' business. Putting them in charge of [spin-off] units gives them that kind of responsibility, and prepares them for broader management responsibilities." Several current members of general management in the company are spin-off venture alumni.

Chapter 3

How Companies Organize Spin-off Units

THE WAY IN WHICH a spin-off operation fits into the firm's organizational structure usually depends on the size and purpose of the operation. In most of the cases studied, spin-off activities have no separate organizational identity of their own. They merely have informal status within existing company units.

Where spin-off operations are instead set up formally, as identifiable units, the most common arrangements are:

- *As a profit center within one of the company's regular organizational units.* The spin-off unit is often established as an appendage of an existing department, with costs and revenues from the venture segregated as befits a profit center.
- *As a separate operating unit.* Some firms have chosen to set apart their spin-off ventures as separate organizational entities. In such a case, a spin-off group built around a manufactured product may have its own production and marketing capabilities. Like other spin-off units, however, it usually relies on regular company units for accounting, advertising and other traditional staff services that it requires. (A very few spin-off operations have grown to such dimensions that the sponsoring firms have formed them into subsidiary corporations.)

Selecting Personnel

Converting a captive operation into a commercially oriented one has seldom meant a change in the personnel assigned to the operation. Usually, the "founders" of the product or service – or at least key members of the supporting team – were influential in the company's decision to sell to outsiders; and they then became prime candidates to manage the new units.

Where a "second generation" of managers has taken over, they are apt to be much like the original group. A second generation of spin-off group leaders in a chemical company came from the same place as the founders – the company's product development laboratories. According to a vice president of the firm, "The ideal candidate is an entrepreneurial-oriented individual who, while he appreciates

Profiles of Spin-off Leaders

Similar observations from other reporting executives whose companies' spin-off operations are of some consequence give this picture of what an "ideal" group leader might be like:

(1) Interested not only in starting new projects but also in implementing plans and ideas. Companies are wary of individuals who have little taste for follow-up.

(2) Broad-gauged, with a diverse (e.g., technical, managerial, marketing) background. A catholicity of skills is often seen as necessary for handling the multiple responsibilities of managing a substantial spin-off enterprise.

(3) Able to gain cooperation and assistance from other company units and individuals outside the spin-off group.

(4) An inspirational leader who can provide a healthy infusion of team morale and camaraderie within a small, separate piece of the company. (Autocratic leadership is cited by some executives as one of the reasons why spin-off operations in their companies' groups have fallen short of expectations.)

(5) The capacity to select individuals who will, like their leader, be self-starters is also deemed of some import. Since the units are often small, each staff member usually has a major impact on the success or failure of the venture.

the backup of a large corporation, wants to do things on his own. Our spin-off groups have a great deal of freedom. There's no one looking over your shoulder, so these managers have to be goal-directed and 100 percent self-starters."

Motivating the Spin-off Group

Compensation and other motivational incentives applied in most spin-off units are no different from those established for other related elements of the companies studied. Several managements say that extra motivation is not really required. Some are especially lucky in this regard, as in the case of a spin-off unit manager in a steel company who reports that "being in this group itself is a form of corporate recognition. My staff knows that we're the 'Green Berets' of this company and they react accordingly."

But not every spin-off unit executive feels this way. A few express annoyance at what they have found to be a lack of recognition, complaining that their activities are so foreign to the company's normal business concerns that they are looked at with curiosity and as outsiders. A spin-off executive in a food company says, "We're stuck off here in the corner and top management doesn't have the foggiest idea what we do."

Chapter 4

Marketing Spin-off Products and Services

TRADITIONALLY, THE commercial sale of spin-off products or services tends to be, in the words of the chief marketing executive for a machinery producer, "strictly a low-pressure operation." In most situations, this is attributable to:

- The relatively small amount of risk capital required to convert to commerical operation.
- Factors contributing to a "seller's" market.
- Assured purchases by the unit's most important "customer," no matter what happens.
- The absence or avoidance of strong competition.

For many spin-off ventures, nearly all of the investment needed to supply the product or service, even to an expanded "market," has already been made. Thus, there may be no serious question of "risking" funds in the development of a commercial offering; often, what is mainly required is additional money – though probably a modest amount – to promote the venture.

Sometimes the original decision to offer the product or service to outsiders was prompted by the requests of other companies, in which case a certain minimal market base may be assured. Also, unlike other marketers, those serving spin-off units are comforted by the fact that they are unlikely to lose their major "customer" – their own company. For many of the ventures studied, more than half of the output is "sold" to the sponsoring company.

A substantial number of spin-off ventures are relatively free from severe competitive pressures. Some that serve markets which are strictly local, or otherwise narrowly selective, say that competitors are "nearly nonexistent," because of price advantages their units enjoy. Several deliberately avoid head-to-head confrontation where such advantages do not clearly obtain.

Furthermore, marketing is apt to be at a relaxed pace when conducted on a "part-time" basis. "Selling this group's output is only a small part of my overall responsibilities," says a spin-off manager in a furniture company. "Although we're doing a good job with our sales effort, I wouldn't get too excited if sales slacked off." This distinguishes the spin-off venture from the more traditional campaigns by which companies launch their new products.

Test Marketing Spin-off Ventures

The reported test marketing of a spin-off product or service is more often the exception than the rule. One reason why formal, comprehensive market investigations so seldom precede the shift to commercial sale is that the viability of the product or service in meeting the company's own intended needs has already been established.

Nonetheless, a few firms studied did attempt to assess potential market reaction to their spin-off offerings in advance. The market evaluation was generally conducted by the company's own marketing department. A few firms used outside consultants for such analyses. Usually, little attempt was made to mask the identity of the would-be supplier, partly because emulation by competitors was seldom a worry. The research frequently consisted primarily of collecting price data and other information about similar products or services already being sold to prospective buyers.

Perhaps most favored has been the "make a little – sell a little" approach. Two or three likely purchasers are selected as targets. If they are successfully sold, then the sales effort is expanded. Along the way, the company can begin to judge more accurately how the product or service works out for outsiders and what it can expect from the venture.

Pricing Spin-off Products or Services

Seldom has much formal research gone into determining an optimum price for spin-off offerings. Where costs might at least bear on the question, many reporting executives cite difficulties in pinpointing costs. A food company executive states: "We're not supposed to be using a 'built-up' method of product pricing, but costs are the prime determinant in our strategy. With this venture we find it difficult to pin down investment costs, although we do keep tab of our operating costs. So, built-up pricing becomes the ruling method."

In several instances, prices have been arrived at simply by finding out what prospective customers have been paying for similar products or services and then matching this level. In other instances, where the issues were less clear, managers have been forced to guess at what customers would be willing to pay. A few say that this approach resulted in initial underpricing of their offerings. Later, adjustments were made when it was realized that customers were quite willing to pay more.

What is mainly sought, several managers say, are prices reasonable enough to guarantee a steady demand. There are sometimes reported penalties for stimulating too much demand with "bargain" prices. Such was the case with an electroplating service which was priced low – but at a level that covered costs and returned what the management regarded as fair profit. But planners had neglected to examine carefully the prices competitors were charging and also to take fully into account the company's own costs of investment in production equipment. Within six months after the venture's inception, would-be customers from as far as 700 miles away were sending plating work to the company. Independent competitors began protesting that the charges were out of line. Following a quick reevaluation and confirmation of that fact, an adjustment of the pricing structure was made.

On the other hand, overpricing carries a different set of potential penalties. One machine tool company ran into trouble because of misunderstanding about the backup support encompassed in its prices. The introduction of its spin-off, a heat-treating service, was greeted with monumental disinterest by hoped-for customers. Investigation showed that the price being asked was equal to or slightly lower than the prices quoted by competitive suppliers. But many prospective customers, aware that the offering was being made by a spin-off group, were under the impression that the amount of accessory service to be supplied – consultation, testing, etc. – would be less than that available from the independent competitors. Although this assumption was false, potential purchasers had been put off by the belief that the company's heat-treating service was actually "priced higher" than the rest because of a supposed lack in back-up support. To overcome this problem, the company reduced its price and thus gained entry to the market. During the first year, it was able to convince its customers that it was providing adequate support, and thereafter it gradually raised its prices to a level nearer those of its competitors.

Product or Service Design

Under traditional marketing philosophy, it is axiomatic that products or services conform to market need. Spin-off items, however, have been tailored originally to the preferences of their producers. Often, they are offered initially to others on an "as is" basis. Real market acceptance, however, sometimes forces some changes in design.

For example, one firm found that a manufactured product, whose form was suitable for its own use, was bulky and difficult for others to use in their factories. So the product as sold outside had to be redesigned to less than one-half its original size, and simplifications were introduced to facilitate its use by semiskilled labor.

Selling

With few exceptions, spin-off products and services are reportedly sold direct to their users. Formal sales representation is more the exception than the rule. "Selling" is often a part-time activity for key personnel managing the venture.

In other cases, sales representation is by one or two members of the company's regular sales force – possibly on a part-time basis. (For an example, see Interpace Corporation, below.) The spin-off entry may be at a disadvantage if lost as a minor possibility among the salesman's offerings. "We'd like to get our sales effort going on more than a hobbyist basis," one executive complains. "We never get the salesmen sufficiently enthused about our product to establish momentum and make a real breakthrough. We know there must be more people out there interested in our product, but we can't get the salesmen to make the necessary number of calls. In fact, we can't even get them to follow the live leads we provide."

Selling through agents or distributors has proved worthwhile in a few cases where the potential justified such a push and the company's own sales force had little familiarity with or access to the market to be served. Another occasional option is that exercised by a chemical firm (see page 20) which had developed an instrument for use in measuring factory noise conditions. When it elected to market the instrument, it made use of the distribution capabilities of another firm specializing in the manufacture and sale of work-safety equipment. But later, as sales grew, the chemical firm developed its own distribution network.

Where appropriate, reliance on qualified outsiders to carry out the selling job seems to have worked out fairly well.

The Corporate Umbrella

The sponsors of many of the spin-off ventures studied are large firms. Those whose corporate identities are well known throughout the business community say that, generally, this works to the advantage of their spin-off ventures.

In particular, the "halo effect" is strongest when: (1) the relationship between the spin-off item and the sponsor is seen by prospects as clearly relevant and synergistic; (2) the financial stability of the operation would frequently be an important consideration in a prospect's selection of a vendor for such a product or service; and (3) the quality of the product or service – e.g., its reliability or the certainty of its supply – would often weigh heavily in choice of vendors.

There is not always such solid reinforcement of the spin-off's position when the logical linkage between the parent company's principal business and the nature of the spin-off product or service is extremely remote or unclear. One food processor, with a large fleet of company-owned trucks, developed a truck lubrication service. Its mobile units would travel to various truck depots and perform periodic lubrication,

The Territorial Imperative

Interpace Corporation, a producer of concrete pipe and building supply materials, has a spin-off operation that markets a special line of high-tensile, prestressed products, including wire and pipe-joint rings.

Many years ago, Interpace engineers had shown that it made sense for the company to fabricate its own pipe-joint rings and to process the specialized wire used in its production of reenforced pipe. This would assure the company a dependable source of supply for these items. When the necessary production equipment was acquired, it was capable of producing substantially more of the special products than Interpace could use; and so the company began to sell the excess routinely to other pipe producers.

Interpace only markets these special metal products in those territorial areas where it does not produce pipe of its own.[1] The company has approached pipe sellers in those areas, pointing out that because Interpace is not a supplier of pipes in their markets, it is willing to sell its specially developed reenforcing steel and pipe rings to them. This has proved successful, and the spin-off venture has remained viable.

[1] Given the weight and bulk of concrete pipe, manufacturers must have local plants in order to serve nearby construction projects.

a service that other independent vendors also provide. An executive of the company notes that some prospects were put off because they "could never quite figure out what a food company was doing in the lubrication business."

Identification with the sponsoring company has not always proved a boon to spin-off marketers for other reasons. Thus, when prospective customers are already suppliers to the sponsoring company, there may be worries about possible charges of reciprocal trading.

Chapter 5

Spin-off Performance

CONVENTIONAL MEASURES of performance – for units and personnel – are being appropriately and effectively applied in spin-off ventures that are separately maintained and hold profit-center status. Judging worth and accomplishment is often reported to be much harder, however, when such ventures are part of larger operations and where costs and activity measures are not easily segregated.

Thus, companies sometimes find that the true costs of setting up and monitoring product or service facilities – which form the basis for spin-off sales – are buried in other units' budgets or comingled with other categories of costs. Attempts to identify such costs have proved futile for some. Furthermore, initial estimates of potential outside sales and return on investment for many ventures are made relatively informally.

In a metals company that sells a specialized training program to outsiders, the project director recalls, "Management originally asked us to provide return-on-investment projections – before allowing us to market our program. We had a pretty good idea as to how many of the training units we could sell, and thus what our gross revenue would be. But finding out exactly how much we had invested in the program was difficult. We had borrowed staff from several departments in order to develop the program, and many of the production costs had been absorbed or disappeared. The best we could do was to estimate what it might cost to begin such a program from scratch. But these estimates proved so high that trying to apply them in an ROI calculation was ludicrous."

Similarly, Du Pont's safety programs – now sold to other companies – were the result of decades of research in occupational health and safety hazard technology. "There is just no way," asserts a company executive, "that we could put a precise value on what it cost us to develop them."

Attempting to apply regular performance standards to their spin-off businesses would lead them astray, say several managements. For example, one office equipment producer has established 21 percent as the minimum acceptable return on investment for its new ventures. After two years, an internally supported service offered to outsiders was providing only a 10 percent return on investment. Under ordinary circumstances, the venture might have been dropped. But the spin-off service – an inventory control system – had been useful not only in reinforcing the company's sales effort, but also in assisting the company's distributors to be more effective in managing their inventories. Management concluded that, on several counts, the venture was well worth continuing.

In another firm, a spin-off venture was bringing in virtually no return on investment – by standard computation. But without the venture, the company probably would have been forced to discontinue one of its manufacturing operations and to purchase certain components from outside vendors. It was estimated that the resulting increase in its costs would have been about 25 percent.

As suggested, the real payoff for some spin-off businesses is in rewards to their sponsors which – if not entirely "nonfinancial" in the long run – are at least not readily measurable. Such rewards may be reflected, for example, in the experience gained by managers assigned to these ventures. Says a chemical company executive, now a senior vice president, "Managing the spin-off group was one of the most rewarding assignments I've had. It certainly taught me the value of self-reliance and the importance of selecting a good management team."

Spin-off Failures

By whatever relaxed standards it may be judged, not every spin-off business is successful. As might be expected, numerous opportunities arise for making mistakes, once a company moves beyond its normal range of commercial competence.

Perhaps such mistakes would be even more serious — and the failures more numerous and clear-cut — were it not for the special character of so many spin-off ventures. At the beginning, both the scale of their operations and the expectations of their sponsors are often rather modest. And, given the assured patronage of their principal customer, the ultimate penalties for error may tend to remain somewhat limited.

When spin-off businesses fall short of sponsors' hopes, the reasons most often cited are:

- The product or service offered had no real advantage or appeal for its intended prospects;
- The company's planning prior to "going public" was inadequate;
- Mistakes were made in pricing the commercial offering;
- Backup support given the venture by the company was inadequate.

As suggested, sponsors are sometimes overly optimistic in judging the true commercial value or competitive strengths of their market entries. Commenting on the problems encountered by some firms in his industry when marketing technical services originally developed for their own use, the vice president of a chemical company notes that "their prices weren't any better and their services were poorer than customers could obtain from commercial labs, and so a market never developed."

When advance planning is perfunctory, as is often the case, there can be disappointments for the spin-off's sponsor. Such failings, plus an exaggerated view of the edge a product or service will enjoy in the market, have sometimes led to overpricing at the outset and less-than-hoped-for sales.

While some ventures are reported to have thrived with minimal marketing efforts to back them up, it was the lukewarm support given to others that doomed them from the beginning. Says one manager, "It was a matter of 'go ahead and sell the product — as long as it doesn't cost us anything.' "

The personnel vice president of another firm that makes considerable use of spin-off groups believes that those who are assigned as their leaders often seem to become better managers a result. Among other things, he says, they tend to be more innovative and independent.

Those working in spin-off businesses are reported, in still other cases, to be better motivated because of it. The director of an analytical laboratory in a petrochemicals company notes that the turnover rate among his staff, since it became involved in a spin-off venture, is one-half that experienced in other company labs.

Some spin-off groups have also proved valuable to their sponsors in attracting new personnel. Several research luminaries of one firm are said to have first entered the company through its spin-off unit. Later, they gained higher positions in the company's main research and development laboratories.

Occasionally, there are public relations benefits for spin-off sponsors. One executive reports, "The [spin-off venture] is the first really new thing that's happened around this company in more than a decade. It does not have to make a lot of money to be a success. Its simple existence — and the publicity it has attracted — have already made it a winner." And a steel company, hoping to exploit the publicity potential of a new spin-off consulting service for pollution abatement, is eager to "let the world outside know that we're doing something besides making steel."

Part II
Selected Examples of Spin-off Ventures

Cosco Home Products, Inc.

MANY OF THE products of Cosco Home Products, a manufacturer of furniture and home furnishings, require large quantities of molded plastic parts and metal stampings. Long ago, Cosco set up facilities for making such components itself. More recently, the company decided to try to sell some of these parts regularly to outsiders. Cosco had first experimented with the sale of component parts in 1957. At that time, the firm began to sell some to a few local manufacturers, but soon discovered that it lacked sufficient production capacity to meet both its own needs and those of outsiders. Later, the expansion of Cosco's production facilities changed this situation. Cosco eventually established a separate sales organization, which now supplies such parts to other manufacturers in the Indiana-Ohio-Kentucky area.

The Marketing Effort

At present, Cosco's spin-off group is not a separate organizational unit of the company. A sales manager directs the marketing effort of the venture, referred to as "O.E.M." sales, and reports to Cosco's vice president of merchandising. An operating committee, composed of executives representing several different departments of the company, helps provide new product ideas and further operating guidance.

In the beginning, some of the company's regular customers were asked about their possible interest in custom plastic molding or metal stamping. At first, a few of these prospects expressed reservations. A Cosco executive remembers that one buyer, after hearing about Cosco's capacity to produce such items, replied: "I'm not impressed with all those machines. As long as you've got your own product line, it's going to take priority."

This was recognized as a potentially serious objection that had to be overcome through performance and on-time delivery. Management had to make sure that Cosco's manufacturing personnel understood that outside customers for moldings and stampings would have the same priority as those inside the company. Cosco salesmen pleaded with prospective customers for "an opportunity to fail." Fortunately, such failures did not take place; in fact, the spin-off group has recently been awarded a certificate of merit as one of the outstanding vendors in its area.

Company executives believe that Cosco's reputation for quality control and product safety was an important factor in promoting the commercial sale of components and that this made possible the success of the venture. The company's extensive product design and engineering staff, it is believed, may also have attracted potential buyers. Parts customers, given ready access to the company's product design group, were not simply buying a "surplus" product. Rather, they were purchasing a *custom* plastic molding and metal stamping and finishing service.

Calming Internal Fears

At times, managers of the O.E.M. venture had to assuage the fears of some members of the regular sales force that support of the firm's regular product lines – i.e., its assembled products – might be weakened somehow by the new effort to sell components. But, a company executive says, "We planned for such a possibility and made certain that such fears never reached crippling proportions. We also made certain that we didn't get into a situation – or an account – where our regular product line could be weakened."

Analyzing the Benefits

According to Cosco's management, the O.E.M. venture has permitted better utilization of the firm's production facilities and has shortened the payback period on some of its machine investments. During fluctuations in the demand for housing and furniture, component sales by the O.E.M. group have helped in maintaining more stable employment and production.

There were also other benefits. A company officer states that those in charge of the O.E.M. venture have substantial latitude in devising and executing marketing strategy. "Our people found that when they got a new customer or project, they had to stay with it right from the start to the finish. This meant pricing it and getting production out in two weeks, not two years. The experience of meeting other people's deadlines has been eye-opening – and rewarding."

Because experience with this contract-manufacturing venture has been both profitable and useful, Cosco is now seeking quotations utilizing additional facets of its manufacturing operations.

Diamond Shamrock Corporation

THE ENVIRONMENTAL Laboratories unit of Diamond Shamrock Corporation has established the methodology for monitoring the company's conformance to environmental rules and regulations. To carry out its mission, it has developed the specialized analytical capabilities required for trace-metal and organic chemicals detection.

In its work, the laboratory group for several years has been conducting waste-water surveys and air-pollution analyses for the firm's plants. In addition, it performs industrial hygiene surveys in order to assay the company's compliance with OSHA standards.

An Offering to Outsiders

Recently, the company decided to offer the testing services of the Environmental Laboratories to outsiders, and it set up a small sales force to do so. A review of the fees being charged by competitive testing laboratories helped to establish a pricing schedule. Then the company's advertising staff prepared a brochure describing all the services available.

Environmental testing, by its nature, means involvement with governmental regulatory agencies. As a major chemical producer, Diamond Shamrock necessarily keeps in close touch with environmental regulatory agencies at city, state and federal levels. These contacts, and the diversity of Diamond Shamrock's operations, have reportedly helped the laboratory in establishing its credentials in the testing field.

From an organizational standpoint, the Environmental Laboratories unit reports to the company's research center. Although it accumulates expenses and records revenues, it is not organized as a profit center.

Work Priorities

Among the services offered by the spin-off venture are field assistance with sampling, sample pick-up, etc. This requires the assignment of certain staff members to work outside the lab, and the question sometimes arises as to whether priority is to be given jobs performed for the company's divisions or to those of outside clients. "It's constantly a consideration," the lab director states. "But where we decide to make a service public, we're duty-bound to provide the outside customer with the best service possible. I know of some other such ventures that have had priority conflicts – and, in some cases, their outside work received lower priority. In our case, we try to give each job equal status."

Extra Benefits

Diamond Shamrock's management believes that offering the Environmental Lab's services to outsiders has gained the company more than just extra revenue. Says one manager, "Getting out to the public gets you into areas of technology that you would not ordinarily experience. We believe we've obtained much more specific experience and detailed information on relationships between the environment and the chemical industry. Secondly, getting our people into a competitive market has proved to be of real value. There are staff members who prefer to get into the 'business' aspect of our work. Selling to outsiders has provided them with an opportunity to accomplish this and still retain their links to our technology."

Experience with the spin-off venture has prompted Diamond Shamrock to expand its laboratory staff. It is also investigating the possibility of adding other testing services to the "product mix" of its spin-off group.

Applied Technology Division, Fabrics and Finishes Department

E.I. du Pont de Nemours & Co.

THE DU PONT applied technology division was established in the early 1970's with the basic charter to market selected technologies, programs and hardware developed in Du Pont's basic business operations and deemed useful to other industries. This group, which is now part of the fabrics and finishes department, was an outgrowth of an operation in the organic chemicals department.

The division's beginnings were based on a library of vocational and safety instruction courses which were being marketed, primarily as a customer service, to the petroleum industry. In enlarging the training material base, the group found other areas of expertise in demand and, within the first year of operation, initiated commercial consulting services, sold computer programs, and provided industrial hygiene instrumentation.

Du Pont's applied technology division did not attempt to focus their marketing activities on existing corporate customers, but addressed the general industrial community. However, management agrees that with a firm as large as Du Pont the corporate logo is valuable in gaining a customer audience.

Du Pont's experience with CAPS, a computer-aided pipe-sketching program, is illustrative of one of the proprietary products marketed by the applied technology division. With the assistance of computer experts, Du Pont engineers automated the process of drafting pipe sketches by means of a computer system that generated detailed isometric drawings as well as materials purchase requirements, fabrication instructions, cost estimate information and inventory control data.

Such computer programs are costly and time-consuming to develop. The Du Pont program allows the user to get started promptly and includes not only the computer software, but also on-the-job training at a Du Pont facility.

Safety Services and Products

A company spokesman points out, "Safety has been a prime management function since nearly the founding of the company. For the last several decades Du Pont has been regarded as one of the safest companies in the industrial community, with an extremely low accident frequency rate. The safety consulting service is a distillation of this corporate experience."

Du Pont consultants, the company says, are experienced in both the technical and management aspects of work safety and are active at all levels in a customer company. In many instances, work with client companies results in significant reductions in accident frequency rates.

Energy Management Services

Fortuitous timing reportedly played an important role in the market development of another service sold by the applied technology division. Du Pont established a heat, light and power division in the early 1900's and has attempted to manage energy usage ever since.

Business concern over rising energy consumption and costs prompted the applied technology division to offer internally proven programs aimed at cutting energy use. The division now has an energy management services group which shows clients how to achieve energy savings of between 10 and 20 percent on their existing operations.

The company reports that its energy consultants have worked with numerous companies and have made energy surveys of over 200 production facilities. Du Pont engineers state that about 35 percent of the energy savings are achievable with little or no capital investment on the part of client firms.

Industrial Hygiene Instrumentation

Several industrial hygiene topics have also been found of interest to the applied technology division, particularly noise and air pollution. For example, the audio dosimeter, an outgrowth of Du Pont's hearing conservation program, was marketed to outsiders in the early 1970's. This device, which may be worn by industrial workers, measures daily noise exposure and thus identifies workers particularly subject to potential hearing loss.

The audio dosimeter was enthusiastically received by the marketplace, leading the division to introduce several new models of the dosimeter as well as other products, such as personal air sampling devices.

Operation of the Applied Technology Division

Internally, the applied technology division is organized into several groups of specialists, much like a small company might be. There are product managers for the various products and services, a divisional sales force, and a sales manager. Specialty staff services, such as advertising and public relations, are paid for out of the operating budget on an as-needed basis.

Many of the products or services offered by the division have been used at Du Pont for many years. Thus, company executives explain, there is little need to test market the products or services. Even so, each new product or service does require, the executives say, "learning a new business," and sometimes that learning process takes time.

Overall, the division's sales and profits are accurately measured and its cash-flow and operative earnings contribution provide a ready measure of performance for Du Pont's top management.

Extra Benefits

Besides providing additional profits, the applied technology division is also returning some intangible gains to Du Pont. For example, in the case of instrumentation an instrument developed for internal use and then taken to the market often turns out to possess design features that it might not otherwise contain. Improvement in the instrument is made possible both by larger scale production and by the demands of the marketplace.

Looking ahead, Du Pont executives expect the division to expand in both scope and size. For example, the company is currently developing a new training program dealing with the handling of hazardous materials – again based on Du Pont's own expertise – and expects to market this know-how to other companies.

Florida Steel Corporation

TRAINING THE OPERATORS of overhead cranes and forklift trucks has been a persistent problem for Florida Steel Corporation. Often, personnel recruited to these jobs have limited education and little familiarity with machinery.

The manager of human resource development at Florida Steel sought to meet this training need by developing a "learn it yourself" programmed instruction text, along with an audio cassette, for training recruits in the operation of materials-handling equipment.

The training format, as indicated, is largely self-instructional and is followed by a performance checkout on the equipment. It takes approximately one hour and twenty minutes to complete the forklift program and from one to three hours to complete a lesson in the crane operator training program. Crane trainees take as many of the following five lessons as necessary— general crane safety; materials handling, parts I and II; pendant cranes, cab cranes, and hot metal cranes. The program concentrates on the nomenclature, basic principles of operation, and preventive maintenance for lift trucks and/or overhead cranes.

Another feature of the program is that, due to the nature of the self-instructional material, a supervisor need spend only about one-third the time on the program as that spent by the trainee.

The program proved so successful for Florida Steel that the company recently decided to market the program as a spin-off venture. A company executive says, "Nothing in our training program is really proprietary. That is, it doesn't necessarily give us a competitive advantage over other steel producers. So, we saw no reason to restrict the material." Rather than sell the programs directly, Florida Steel entered into an agreement with Westinghouse Learning Corporation to market the programs.

Inside the company, the programmed instruction approach has proved popular for training and has produced a demand for additional training programs. Several other instructional programs are in the works. The company has used outside consultants to aid in the design and production of some of its training courses, but the crane and forklift truck programs are based almost entirely on Florida Steel's own job-training experience.

It is too early to assess what profit might be generated by the spin-off services. "If we can sell these programs, and make some money on them," says a company spokesman, "we'll be paying off our own training investments."

General Electric Company

ONE LARGE FIRM that markets several spin-off products and services is the General Electric Company. Illustrative are three internally developed services which include an economic forecasting service, a management development program, and a technical service whose subscribers gain access to abstracts of published articles from various sources. Out of a half-dozen or more spin-off ventures that GE has, these three provide some idea of the typical rationale, as well as the organization and operational options, for such ventures.

Economic Forecasting

Several years ago, an expansion of GE's corporate planning effort made it necessary to develop a broader data base. As part of this effort, GE economists developed a computer-based econometric model intended primarily to fit the company's own forecasting needs. At the same time, the model was made available on the company's time-sharing facilities.

By the late 1960's, the planning department's forecasting services were mainly for the benefit of GE's chief executive officer. The group would meet with him quarterly, offering its analysis of the company's future prospects by major product or industry components. The chief executive valued this input and was among those who favored making the underlying economic projections and assumptions available to outsiders. Not all company executives were happy with the idea. Some were concerned, for example, that if economic forecasts went wrong GE customers might be hurt and would blame General Electric for it. But such fears proved illusory. And a company spokesman says that having an objective format available for GE customers outweighed other objections.

Reorganizing the Unit

The economic forecasting operation had begun as a staff element of the planning department. It was then decided that it should be set up as a separate profit center, in another part of the company, so as to facilitate the marketing of its service and the appraising of its performance.

However, this arrangement produced some difficulties. The forecasters no longer had their previous close access to the corporate staff or executive officers. And they often lost time as they became enmeshed in the nuts and bolts of costing, pricing and marketing their services. After reexamination by management, the forecasting group was returned to a staff function, and responsibility for the marketing of its services was transferred to the company unit that sells GE's computer time-sharing services.

Much of the staff for the enlarged economic forecasting group was recruited from outside the company. Management went looking for economic "generalists" to assist in developing several different "products" for the unit.

Users of what is called MAPCAST – Management Analysis and Project Forecast Service – have access to GE forecasts via teleprinter connections. They can retrieve component data back to 1950. In essence, MAPCAST subscribers receive a business environment forecasting service, of the same type that GE forecasters prepare for the firm's own management.

Marketing Considerations

It was believed that the GE name would help to sell the MAPCAST service and in general this proved to be valid.

Deciding on a price formula for the forecasting service turned out to be difficult. Originally, the service purchased "time" wholesale from its own computer services, marked it up, added the costs of the forecasting operation, and then sold the entire package. But now the unit pays "full rate" for such services.

At first, pricing was influenced largely by the costs that customers could be expected to incur if they were to attempt to replicate such a service on their own. Later, however, strategy came to be based more directly on market considerations – what prospects would have to pay for competitive services and what they would be willing to pay for the GE forecasts.

Initial marketing efforts were aided by marketers' access to an excellent prospect list of existing GE customers. In addition, the staff of the forecasting unit had been active for some years in an informal consortium of New York-area economists who operated a cooperative data bank. Some of these other economists' companies became customers for the new service.

While the forecasting group was still doing its own marketing, some of the staff members functioned as programmers in the morning and as salesmen in the afternoon. Reportedly, they "loved it" and came to know their "product line" and capabilities much better as a result. At present, however, the sale of the forecasting service is the responsibility of GE's information services division. The effort consists of personal sales calls, media advertisements, and direct-mail campaigns.

Evaluating Performance

The forecasting service, which no longer operates as a profit center, does reserve some of the revenues obtained from the sale of its services. This revenue is increasing and serves as one barometer of performance. It also allows the group to maintain a larger staff of specialists than might otherwise be possible. Company executives say, however, that since the service continues to serve GE's own economic forecasting needs, any appraisal of its worth must take this into account.

GE Management Development Institute

Since its founding in 1892, General Electric has generally favored a "grow-your-own" philosophy in building professional manpower. Its college recruitment and managerial development program is designed to provide a continuing source of new candidates for management positions in a company that has over 75,000 exempt employees – including 40,000 college graduates.

These efforts are centered in GE's Management Development Institute, a training facility occupying a 30-acre campus at Crotonville, New York, and capable of handling 35 full-time students. The Institute is a component of the professional development operation, which is a part of the corporate employee relations staff. The company has 15 regional development centers managed by various GE operating components.

The Institute relies on a full range of training techniques – lectures, classroom discussions, case studies, seminars, audiovisuals, programmed learning, business-game simulations, and so forth.

Opening the Doors to Outsiders

In 1970 it was decided to sell selected GE training expertise to other companies. There were several reasons for this decision. First, a change in accounting practices required that the Management Development Institute be able to liquidate all of its costs. Second, it was believed that the "discipline" of the marketplace would help assure reliance and quality.

GE's senior management was interested in broadening the scope of the Institute's programs. Selling its services to outsiders would help to achieve that goal. Many of the instructional programs offered by the Management Development Institute were not proprietary, and GE management judged that it might be advantageous to have the company's customers benefit from them. In addition, the company reports, the Institute staff benefits from its interaction with outside students.

Designing a Curriculum

From the many courses developed for GE's internal purposes, the Institute's staff picked seven programs for sale to outsiders. These were the Management Practices Course, Interpersonal Communications Shop, Career Planning, Motivation Workshop, Foreman Supervisory Program, Manufacturing Studies Course, and GESIMTEL – a computer simulation program for managers and supervisors that used a telephone link to a GE computer.

Little in the way of additional instructional staff was required in order to offer these services initially to outsiders. For some courses, however, the training faculty was bolstered by the acquisition of retired executives or university professors who became adjunct faculty members of the Institute's staff. No other major changes were needed.

Management of the Institute says that it is most interested in covering the costs of educational operations – although any incremental revenues are welcomed. From GE's standpoint, allowing other companies to use the Institute has been valuable, not only in covering costs, but also in stimulating enrichment of the content of programs offered to GE's own employees.

Technology Marketing Operation

The Technology Marketing Operation markets a number of publications developed internally by General Electric. In one typical example, General Electric management had recognized that engineers, product development specialists, and other technicians might well miss, because of reading-time limitations, pertinent articles appearing in an ever-increasing number of business and professional magazines and journals. About 15 years ago it was decided to select 100 or more important publications and to abstract articles from them dealing with manufacturing and productivity-oriented subjects. These were combined in a presentation of approximately 150 abstracts by the Technical Information Periodicals Service (TIPS). Later, this publication was offered to other companies.

A company official observes, "We were sort of looking around at some of the publications that were being prepared internally – purchasing manuals or engineering specification works, for example. Our goal was to find technical information that was already being generated in a form that didn't need to be censored."

The operation purchases, in bulk, reprints of articles it believes will be of probable interest to subscribers. The unit is thereby able to mail reprint orders directly, generally within 48 hours. For an article that is likely to be of widespread interest, General Electric negotiates with the publisher the right to reprint the article itself.

The Technology Marketing Operation functions as a profit center and runs in the black. Its offerings are priced, according to a company executive, "high enough to discourage the merely curious – but low enough to encourage a steadily growing list of subscribers."

In addition to the TIPS service, the Technology Marketing Operation also sells heat-transfer and fluid-flow data books, drafting manuals, components technology and standardization manuals, computer software, and periodicals such as *Selected Business Ventures,* as well as supplements for updating various technical manuals.

Kimberly-Clark Corporation

KIMBERLY-CLARK CORPORATION stations its aircraft fleet at the Appleton, Wisconsin, airport. To keep its planes in good working order, the company has always hired its own maintenance personnel. The idea of operating a fixed-base aircraft maintenance function had been discussed with the view that it would partially offset the mounting costs of the corporate fleet. The problem was space in which to perform this maintenance function and, in 1969, this was resolved by the construction of appropriate hangar space. Soon the company decided to organize a separate corporation, a wholly owned subsidiary of Kimberly-Clark Corporation.

The resulting spin-off venture was an aircraft service organization designated K-C Aviation, Inc. By now, K-C Aviation offers one of the largest and best-equipped corporate aircraft repair centers in the Midwest. The venture has its headquarters in a $1 million hangar and repair facility at the Appleton airport. It provides both parts and repair services for jet and propeller-driven aircraft, and its station is FAA-approved.

Aircraft manufacturers, upon learning there was a reliable repair and retrofit center in the Appleton area, appointed K-C Aviation as an approved service center for their aircraft. The business grew rapidly without the need for a formal marketing effort.

Even today, the subsidiary does not have a major problem in selling its services. While sales contacts are made by the sales and service managers, the unit is proud of the fact that a considerable volume of its sales has resulted from the recommendations of its customers.

Organizational Relationship

As an operating subsidiary, K-C Aviation has its own staff. But it does use certain corporate staff services such as auditing, legal and certain personnel services. With a few exceptions, it does nearly all of its own purchasing. For services supplied, the corporate department charges K-C Aviation in the same manner that it charges other profit centers within the company for its services.

The subsidiary could now exist without parental support. It is reported that even if the company should decide to sell its aircraft fleet, it would certainly still retain K-C Aviation, which has proved to be a viable and profitable business entity.

Kimtech Limited

Successful experience with the K-C Aviation venture has led to another spin-off subsidiary, Kimtech Ltd.

The idea for establishing Kimtech originated with the chief executive officer of Kimberly-Clark Corporation. He believed it important to the corporation's future that the finest engineering services be available; however, it was not practical to maintain a complete engineering staff as a continuing part of corporate overhead. To achieve full utilization and efficiency, an engineering organization needs the opportunity to serve many customers.

In early 1974 Kimtech Ltd. was incorporated, a wholly owned subsidiary of Kimberly-Clark. An objective of this subsidiary is to sell services not only to the parent company but to companies in various industries. Kimtech's sales to other companies provide growth in revenue and profit to the parent.

There are two operating divisions. The service division has 240 employees engaged in the design fabrication and assembly of light machinery, and in field construction. The service division offers construction management help in addition to its capability to design and manufacture machinery. The engineering division has 50 employees providing engineering services to the pulp and paper industry, and it includes an analytical laboratory, an environmental laboratory, and a modern pulp laboratory/pilot plant.

Kimtech services are being marketed throughout the world to companies in various industries. Major projects are under way in the United States and in foreign countries. An objective is to develop a broad customer base and thereby minimize the effect on Kimtech of cyclic fluctuations in any one industry or customer.

The engineering division offers complete engineering and design services in the fields of pulp, paper and tissue mills, water effluent plants, and steam plants. Years of experience with capital projects, operations and trouble-shooting have prepared its engineers to take on challenging projects. Economic evaluations, feasibility studies, and design reports are prepared by the engineers to provide clients the information necessary for making decisions about new, expanded or improved facilities or about operating problems. The division's analytical chemistry laboratory has the professional staff and equipment required to provide chemical analyses quickly and economically. Separating and identifying the elements of materials is often essential to solve process problems, correct the cause of product failure, evaluate competitive products, and comply with government regulations.

At the service division there are machine, trade and assembly shop facilities to produce converting machinery for pulp, paper, tissue and textiles. Historically, the majority of these machines have been of proprietary design for converting K-C consumer products such as Kotex sanitary napkins, Kotex tampons, and Kleenex facial tissues.

Obviously Kimtech does not make available to others any customer's proprietary designs. However, the machine design staff is ready to provide designs suited to individual customer needs. For converting and specialty machines, Kimtech provides complete turnkey service, including design, manufacture, installation and start-up. Since each of Kimtech's customers has different machinery needs, standard designs are not satisfactory.

In addition to general services, Kimtech's job shop services include: maintenance service for emergency and scheduled repairs; dynamic balancing of rolls; and design and construction of instrument panels, electronic, pneumatic and hydraulic controls.

Until recently, Kimtech's marketing efforts were primarily personal contacts by departmental managers. More active selling is now required with full-time sales personnel. In the first full year of operation, Kimtech achieved sales of $9 million and was profitable. At this point, 90 percent of sales are to divisions and subsidiaries of Kimberly-Clark. Kimtech's marketing plan is to expand with a growing number of new customers so that, in the long term, sales to non-K-C customers will be half of total sales.

Olin Corporation

NEARLY TWO DECADES ago, Olin Corporation established an environmental hygiene and toxicology department to assay and monitor environmental work factors in the company's manufacturing operations. The company staffed the department with physicians, industrial hygienists, toxicologists and work-safety professionals.

Over the years, this group expanded the scope of its operations to include a variety of environmental hygiene studies. When the Occupational Safety and Health Act was passed in 1970, the group was in position to assist Olin management in providing a safe working environment and to generate data on potentially hazardous conditions or chemicals that Olin employees might encounter in their work.

Many other companies were not as well prepared as Olin. To make certain that their operations conformed to the OSHA standards, these firms often had to obtain the services of outside consultants. A number of them, learning of Olin's experience in this field, made inquiries of the company. These promising signs eventually led Olin to establish an environmental hygiene and toxicology consulting service, built around personnel from its own medical, environmental and hygiene toxicology departments.

Just to meet its own needs, Olin would have to have qualified physicians and toxicologists on its staff. But the workload for such professionals varies substantially. No additional staff had to be hired for the venture and the company could readily control the size of its outside work.

At present, the consulting group concentrates primarily on investigating the impact of various industrial chemicals, dusts, etc., on workers' health. Its members visit client companies and make examinations of workers' blood, urine, tissue, breath and hair. The spin-off services of Olin's group also include noise surveys and evaluation of possible occupational hazards that might be encountered as a result of new processes or operations. The group maintains a toxicology library and directory to assist its clients in evaluating toxicological information and in planning toxicity test programs.

Clients have the option of acquiring the department's services on either a project or annual retainer basis. The latter provides for regular surveys and a continuing program of environmental hygiene control.

The consulting group is not really set up as a profit center. Nor, a spokesman reports, is it highly profit oriented. It does not use national advertising, but relies on word-of-mouth publicity and direct-mail promotional letters to nearby firms or to others in similar lines of business.

H.K. Porter Company, Inc.

DURING THE EARLY 1960's, H.K. Porter Company, Inc., acquired several smaller companies. As did many existing divisions of the parent company, each of these acquisitions had its own advertising and public relations department. On analysis, it seemed that there was little continuity or similarity among the advertising efforts launched by the various divisions, groups and subsidiaries of the company.

Dissatisfied with this approach, top management decided to consolidate all Porter's advertising activities. Advertising personnel from scattered locations were brought together to form a marketing services department. The staff was trimmed somewhat, and unified advertising approaches established.

Shortly after this consolidation, however, Porter sold some of the product lines and manufacturing groups that it had earlier acquired. As a result, the corporate advertising group found itself with less work to do than before – i.e., there was surplus "production" capacity. Some of the firm's magazine advertising was still being designed and placed by an outside advertising agency, although the agency's assignments had dwindled severely because most of the work was being done inside – by what had become an in-house agency. During this same period, Porter established a printing and mailing concern, a move that supported industrial advertising efforts which made heavy use of direct-mail promotion.

When the outside advertising agency resigned the account, this left the in-house group with the total responsibility for Porter's advertising. Within a short time, company management came to the conclusion that since the in-house unit was already serving diverse "clients" – i.e., numerous product lines and divisions of Porter – it might make sense to offer its services to outsiders as well. And so the advertising group was incorporated as Garland Way Advertising, Inc. (Garland Way is the name of an alley behind the H. K. Porter building in Pittsburgh.)

From the outset, it is reported, the unit's affiliation with Porter has been an important factor in establishing credibility. While Garland Way does not "trade on" the parent company affiliation, potential clients are generally aware of the connection.

Selling advertising, printing and mailing services to other advertisers has not meant much of an increase in revenue for Porter. But, according to company officials, it has resulted in improvement in the advertising generated in support of the company's own products. The challenge and stimulation that come with serving outsiders are said to have freshened the thinking and spurred the creativity of the Porter subsidiary.

Ralston Purina Company

IN THE 1960's, Ralston Purina Company began to experiment with a new kind of job-enrichment and productivity-improvement program. The new program, an innovative idea called Operations Improvement, focused on practical, everyday changes that workers and supervisors could institute on their own and thereby have an impact on company profits.

The program was based on the premise that "no one knows more about your job than you do," and its dual goals of increased profits and personal job enrichment were openly linked.

Since the inauguration of the program, more than 3,000 Ralston Purina employees have participated in Operations Improvement. Its success – and news of it – led to the establishment of a spin-off group to provide similar Operations Improvement programs for other firms.

Much of the impetus for taking the venture to the public came as a result of inquiries from other companies seeking to learn more about job-enrichment techniques. In time, 50 to 100 such inquiries were received.

Within a short time, Ralston Purina established two separate Operations Improvement teams, one to work outside the company and a second to administer the program within the firm.

How the Program Works

A typical Operations Improvement exercise begins with a small group of employees meeting for a total of 40 hours during a five- to six-week period. Shown the need for improving profits and eliminating waste, participants are also briefed on various management skills. Next, they learn to identify problems affecting their work areas, to become their own consultants, and to find solutions.

In Ralston Purina's experience, this approach pays special dividends. For example, one of its mailroom supervisors, during an OI program, developed a plan that saves the company more than $120,000 yearly.

Each participating employee's action plan proposes a redesign of his or her work – or that of the department – in such a way as to cut costs and increase personal or departmental effectiveness. Emphasis is on implementing action plans that do not require either major capital investments or employee layoffs – two things likely to encounter resistance. It might be thought that, with these two options removed, relatively meager profit improvements would be possible. But Ralston Purina and its OI clients have reportedly found this not to be the case. Here, for example, are case histories cited by OI management:

- Two major food processors achieved productivity improvements worth more than $1 million with Operations Improvement.
- A chemical company enjoyed savings of $333,000 during the first year of its OI program.
- A meat packer recorded audited savings of more than $150,000 during the first year of its OI program.
- A banking institution, which had enrolled 100 of its employees in OI, had $350,000 savings during the first year as a result.

OI specialists acknowledge that Operations Improvement is no panacea for all employee motivation ills; not everyone responds favorably to an environment of increased personal freedom and self-management. This has to be taken into account, the executives say, in making use of OI programs.

Sybron Corporation

YEARS BEFORE THE current widespread interest in dental health-care insurance plans – and before such plans became an element in contract negotiations – employees of Sybron Corporation already had a company-administered and company-funded dental benefits plan. The Sybron plan, initiated in 1965, was among the first in the country.

One of Sybron's major lines of business is health products, especially medical and dental equipment and dental sundries. It was a logical area of benefits for the corporation and, as Sybron grew, its dental health-care plan evolved and became part of the benefits for employees in other market areas of the corporation. Throughout the United States and Canada, more than 9,000 employees at 25 Sybron divisions are covered by the plan. Coverage also is extended to employees' eligible dependents.

Because of its long experience in administering the dental health benefits plan and because of widespread acceptance of the plan among dentists, in 1975 Sybron formed Sybron Dental Administration Inc. to make the Sybron expertise available to other companies by providing the service of administering employer-funded dental benefits plans.

In developing the dental plan for Sybron employees, Sybron management examined costs as well as the range of benefits, and it decided to "go self-insured." It cited four principal reasons for doing so. First, the overall costs would be less than for coverage obtained through commercial insurers. Second, the company would not have to pay insurance premiums in advance of claims. Third, there would be no need to maintain specific financial reserves. And, fourth, management believed that a company-administered plan would yield special benefits in terms of employee morale and rapport. On this last point, a company executive explains: "It's a rather impersonal process when your medical claims are paid by someone else. You're not as conscious that the company is paying the insurance premiums – or at least part of them."

The staff of the personnel department was enthusiastic about the prospects of developing a company-administered dental health-care plan. The members saw a plus, for example, in that employees' questions about benefits could be answered directly by the personnel office.

Automation

At the beginning, the company processed all dental claims by hand. But once the Internal Revenue Service began requiring all health benefit providers to report payments in excess of $600, manual processing became so time-consuming and cumbersome that Sybron switched to a computer-based claims system. Sybron continued to refine its dental care program in other ways, and the staff expects continued modifications, according to changing conditions.

The program allows employees freedom of choice in their dentists and in the types of dental care obtained. It asks the insured employee (or family) to pay the first $25 of dental bills incurred each year. Then, Sybron pays 75 percent of the balance of all charges. A family covered under the plan for three years could receive up to $1,000 per year in dental health-care payments.

Spreading the Word

Word of Sybron's plan spread among labor and management professionals in the local area. Soon, another manufacturer, contemplating the adoption of a similar dental health-care program, asked Sybron for information about its approach. Members of the personnel department had put together a slide show, detailing how the plan worked, and the actuarial experience with it. They visited the offices of the inquiring company and presented their information.

Later, Sybron was asked whether or not it would be interested in administering the other company's dental health plan. Since Sybron's own plan was already standardized and automated, and since the other company's plan was structured to resemble it, Sybron's management approved the arrangement.

In time, other companies in the state learned of the plan. Several were signed up, and Sybron formed a special unit to administer the dental health-care plans of client employers. The corporation has now formed Sybron Dental Administration Inc. specifically to offer administration of employer-funded dental plans for outside companies.

As additional organizations become clients, and as revenues generated from the management of their plans grow, Sybron expects Sybron Dental Administration Inc. to become a self-sustaining operation.

Jervis B. Webb Company

THE JERVIS B. WEBB COMPANY, a diversified producer of materials handling and production equipment, has many spin-off groups. These include an electrical contracting firm, a control panel fabricator, and an advertising agency, among others. Most of these specialized operations are set up as profit centers.

In-house Advertising Services

Several years ago, the Webb Company relied on a small advertising agency in the Detroit area. The firm's requirements were relatively modest, and the agency's activities generally limited to placing occasional ads and designing direct-mail brochures and catalog sheets.

As the Webb Company grew, marketing management became dissatisfied with the outside agency. The agency seemed to have high personnel turnover, so that Webb executives were forced to spend too much time educating agency staff members about the intricacies of the company's product line. Also, Webb began to expand its own marketing services staff, doing more and more of the ad-related art and photography that might otherwise have been assigned to the agency.

Within a few years, Webb's marketing services group had begun to write and place ads as well. The unit by now had assembled the staff and facilities required for a full-service advertising agency. The decision to incorporate the group as a separate subsidiary – under the title Campbell, Henry & Calvin, Inc. – came quite easily, according to Webb management.[3]

The new company agency began by serving only the 23 operating divisions of the Webb Company. For several reasons, it was decided to take on outside clients as well. One was a desire to increase the size of the agency, and revenues from outside clients could finance this expansion.

[3] For a while, the agency used another name – Ad Venture, Inc. The agency's name has since been changed to Campbell, Henry & Calvin, Inc.

Growing Pains

Some difficulties were encountered at first, but these worked themselves out within the first year of incorporation. On the assumption that the agency was now strong enough to make it on its own, it was moved to outside offices.

Unfortunately, the new venture had been born in the midst of a recession – one having a particularly devastating impact on advertising business in the Detroit area. Competitive agencies were soon scrambling to get a new business and providing existing clients with extra attention. The agency was subject to cost pressures at this stage, and it was decided that it would be more economical for the agency to operate within the physical confines of the Webb organization.

Nevertheless, from the standpoint of the Webb Company, the spin-off still made sense on several counts. For example, the company could draw easily on internal resources to support the unit. Thus, access to Webb's own printing production facilities is an asset when the advertising staff needs a quick color match or a rush printing job. The company's technical artists assist in preparing drawings for advertising of the group's clients. Webb's staff photographer and studio also prove useful to the agency.

The agency's management acknowledges that being part of a major industrial enterprise poses potential problems, as well as opportunities.[4] For example, would-be suppliers to the Webb organization might employ the services of the advertising agency in the hope of gaining some sort of reciprocal trading advan-

[4] Still, protective corporate sponsorship can make life a little simpler. As a competitor of the Webb agency suggests: "With this type of operation, there's not the absolute pressure for survival. They don't have to grab at every possibility that comes along. They can pass up those clients that are tough to work with. That's a luxury independent agencies can't afford."

tage with Webb. Such thoughts are discouraged by stressing to prospective clients that the venture is organizationally divorced from Webb and has no influence over the choice of Webb's suppliers.

Conflicts of interest with respect to job priorities and scheduling could also cause problems if Webb were to get preferential treatment. However, an executive of the agency says: "Webb realizes that for the agency to grow and make a profit, we have to serve our customers in the same way that Webb serves its customers."

In the opinion of one senior member of the agency's staff, persons employed by in-house agencies sometimes get "jaded" working on advertising for the same kinds of products and markets all the time. But with outside clients as well, the staff gets a much broader exposure. "This," he asserts, "helps keep the creative juices flowing" – another reported intangible benefit from the spin-off arrangement.